YOUR KNOWLEDGE HAS VALUE

- We will publish your bachelor's and master's thesis, essays and papers

- Your own eBook and book - sold worldwide in all relevant shops

- Earn money with each sale

Upload your text at www.GRIN.com and publish for free

Bibliographic information published by the German National Library:

The German National Library lists this publication in the National Bibliography; detailed bibliographic data are available on the Internet at http://dnb.dnb.de .

Imprint:

Copyright © 2015 GRIN Verlag, Open Publishing GmbH
Print and binding: Books on Demand GmbH, Norderstedt Germany
ISBN: 9783668255661

Manuel Langer

Aus der Reihe: e-fellows.net stipendiaten-wissen

e-fellows.net (Hrsg.)

Band 1892

Microbal Degradation of Tauropine. An investigation

GRIN Publishing

Intensive Course
"Chemical Ecology"

Subject:

Microbial Degradation of Tauropine

October 18, 2015

Table of Contents

Introduction

Opines

Tauropine, which is a C_5-amino sulfonate (**Figure 1**), belongs to the group of opines. Opines are molecules, that are typically formed in a reductive condensation reaction of an α-keto acid with an L-amino acid.

The first opine to be isolated was D-octopine, which was found in sepia in 1927[1]. D-octopine probably serves as functional analog of lactate for fast energy supply in cephalopods[2].

Opines in the past have mainly been known to play a role in marine animals, such as in slugs, shells, and worms. Under anoxic circumstances, these phylogenically lower invertebrates would exceedingly use an opine dehydrogenase system than the lactate dehydrogenase in anaerobic glycolysis[3]. As this reaction is essential for adjustment of redox levels in cells, opines occur as end products in the anaerobic metabolism.

The degradation pathway of tauropine in marine invertebrates is well known

The product of the reaction of pyruvate and taurine has been called tauropine. Equivalent names for tauropine are D-rhodoic acid or *N*-(D-1-Carboxyethyl)-taurine.

For tauropine, not only its significance for the anaerobic metabolism in various marine invertebrate phyla[3] has been clarified. Scientists also succeeded several years ago in isolating the underlying enzyme, called tauropine dehydrogenase.

The tauropine dehydrogenase is responsible for the reductive condensation of taurine (a C_2 sulfonate) and pyruvate to tauropine. Thereby NAD^+ is involved as electron acceptor. The tauropine dehydrogenase also catalyzes the backward reaction.

This enzyme occurs widely among different animal and plant phyla. For example, it could be isolated from the adductor muscles of *Haliotis lamellosa*[4], *Arabella iricolor*[5], *Asterina pectinifera*[6], *Halichondria japonica*[7], and even from a red alga, *Rhodoglossum japonicum*[8].

The metabolism of tauropine in microorganisms is not yet clarified

Tauropine, besides other opines, has also been reported in the context of bacteria. In fact, it was found in plants, which were infected by agrobacteria with a virulent Ti plasmid[9]. The resulting genetic modification leads to tumor formation, and the plant is triggered to produce opines. As plants cannot use opines themselves, the opines serve as nutrition for the agrobacteria[10] and other opine-degrading bacterial strains[11].

But so far, compared to marine animal phyla, the intermediate steps in the degradation of tauropine in microorganisms are widely unknown. Preliminary investigation in marine bacteria like *Ruegeria pomeroyi DSS-3* and *Roseovarius nubinhibens ISM* has shown, that they can use tauropine as source of carbon and nitrogen. Sulfate thereby occurs as end product.

It is possible, that the tauropine degradation in bacteria is analogous to that in invertebrates. This would mean, that a dehydrogenase is involved. If in

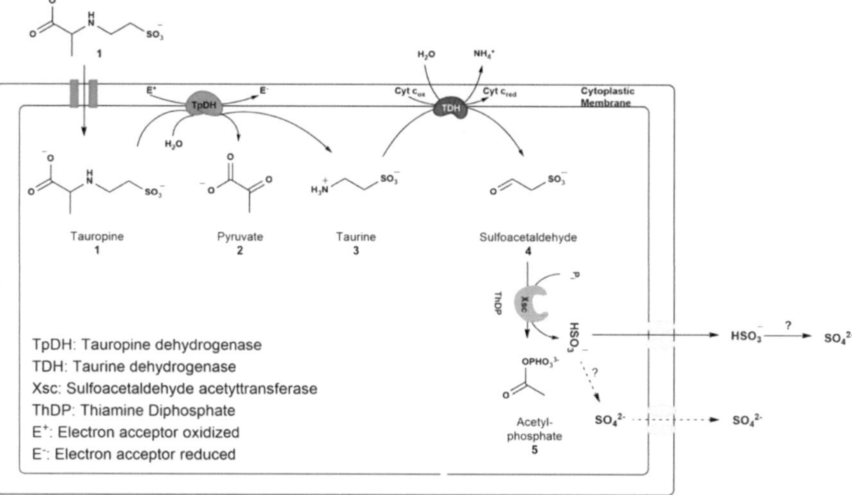

Figure 1: *Potential degradation pathway of tauropine. Inner rectangle refers to the cytosol; the area outside the rectangles refers to the periplasm. Abbreviations: cyt c_{ox} = cytochrome c oxidized; cyt c_{red} = cytochrome c reduced; Pi = pyrophosphate.*

microorganisms tauropine can be degraded into pyruvate and taurine by a tauropine dehydrogenase, it is also possible, that taurine is further metabolized in processes, which are already quite well understood[12-14]. Those processes could include the taurine dehydrogenase and desulfonation by sulfoacetaldehyde acetyltransferase (**Figure 1**).

To conclude, initially one interesting aspect is the involvement and the relevance of one sole enzyme in the microbial tauropine degradation pathway: the tauropine dehydrogenase.

Therefore three main questions were studied. The first was to verify the action of a tauropine dehydrogenase in microorganisms. The second step was to further characterize this enzyme by its molecular weight and its localization within bacterial cells. In addition, the degradation pathway downstream of the potential tauropine dehydrogenase should be clarified.

Therefore, in this study, the metabolism of tauropine in four different model organisms was investigated. As model organisms a *Ralstonia* strain from fresh water was used and in addition three terrestrial bacterial strains were isolated.

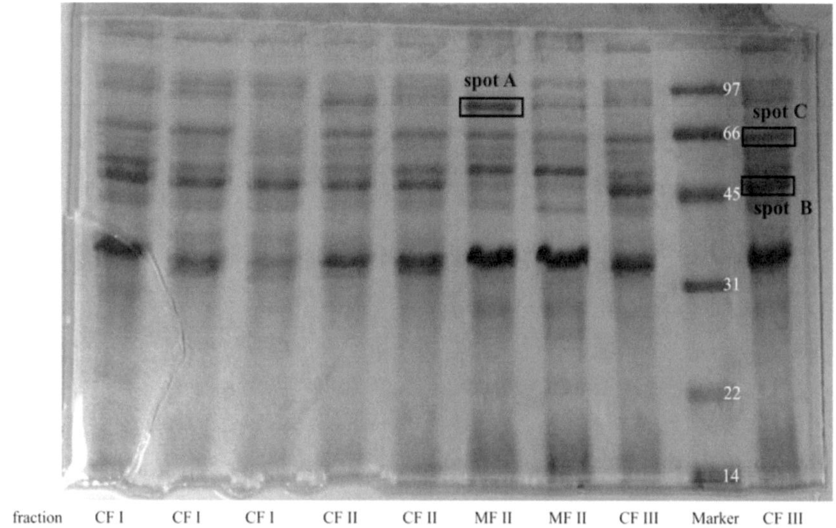

fraction	CF I	CF I	CF I	CF II	CF II	MF II	MF II	CF III	Marker	CF III
substrate	Tp	Tau	Ac⁻	Tp	Tau	Tp	Tau	Tp		Tp
act. [mU/mg]	2.3	0	0	2.9	0	3.9	0	4.7		4.7

Figure 4*: 1D SDS-PAGE of different cultures and fractions. Lane 1-8 and 10: cell lysates, lane 9: marker. Abbreviations: CF = cell-free fraction; MF = membrane fraction; Tp = tauropine; Tau = taurine; Ac⁻ = acetate; act = activity of tauropine dehydrogenase. Roman numbers indicate replicates from different days. White numbers indicate the molecular weight of the protein marker in kDa. Black rectangles define the spots, which were cut out and submitted to peptide fingerprint mass spectrometry (PF-MS) for identification.*

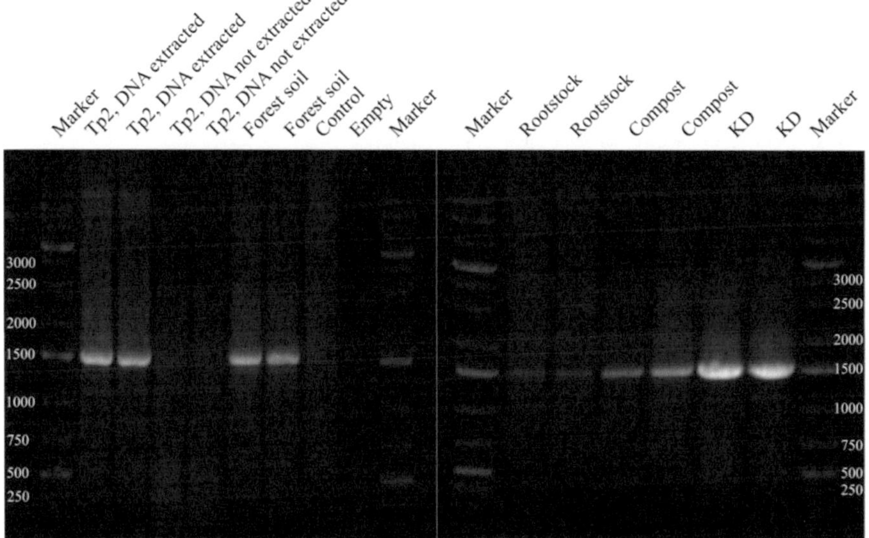

Figure 5*: Agarose gel of the isolated 16S rDNA of the different bacteria strains after PCR reaction, stained with ethidium bromide. The size of the marker bands is written in white numbers and indicates number of base pairs. The captions of each lane are written directly above (strain "KD" was not further investigated in this research).*

Results and Discussion

Based on previous findings in different phyla[3] and the well known degradation pathway of taurine in bacteria[14], a potential dissimilation pathway of tauropine was set up, as depicted in **Figure 1**. For investigations of this pathway, a previously selected strain *Ralstonia solanacearum* was used as a model organism. From previous experiments it is proven, that this strain is tauropine-degrading.

To prove this pathway, different experiments were conducted to validate the presence and action of the tauropine dehydrogenase. Additionally, the presence of the enzymes downstream of the tauropine dehydrogenase, namely the taurine dehydrogenase, sulfoacetaldehyde acetyltransferase, and sulfite dehydrogenase (not shown in **Figure 1**) as well as the end products (sulfate ions and ammonium ions) were investigated. Besides verifying the degradation pathway, new tauropine degrading terrestrial microorganisms were isolated and identified.

Ralstonia solanacearum was cultured in liquid medium, containing tauropine. **Figure 2** shows the growth curve. The first proof of the previously shown pathway (**Figure 1**) is the concentration increase of the metabolites sulfate and ammonium and the decrease of tauropine. **Figure 3** shows the concentration levels of the substrate tauropine, and the metabolites sulfate and ammonium as function of the optical density, which correlates with the time for growth. Although a one-to-one conversion of tauropine to sulfate was assumed, the amount of degraded tauropine is higher than the resulting sulfate ions. The detection method used was not a high resolution method, leading to this small deviation. The level of sulfate is higher than the ammonium level, because nitrogen is partially incorporated into the bacteria.

These findings support the proposed dissimilation pathway.

For detailed investigation of the degradation pathway, the individual enzymes were tested in photometric assays. The wavelength was fixed in each assay and was dependent on the conditions used (for details see section "Methods"). The general approach is the administration of the enzyme (dissolved in a buffer), the respective substrates, a chromophore as a marker, and an electron acceptor. The solution changes its color intensity upon the reaction. Only tests with linear incline or regression of the absorption were considered as positive.

For these assays, three different bacteria cultures were grown using tauropine, taurine, and acetate as carbon source. From these cultures a cell-free, a soluble, and a membrane fraction were prepared, respectively. Roman numerals (**Table 1**) indicate replicates from different days. For validation of enzyme activity, substrate linearity and in single cases the back-

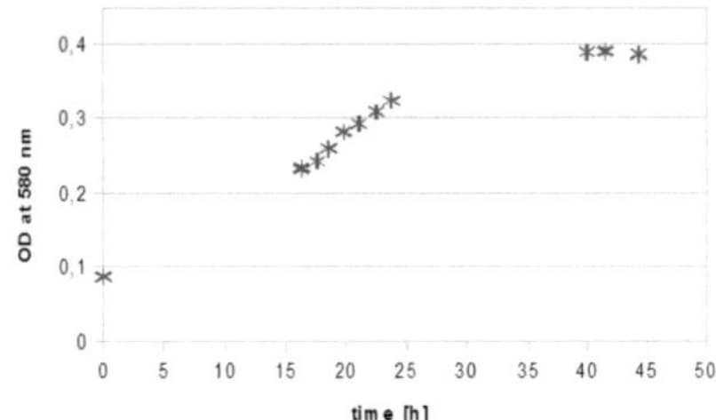

Figure 2: *Growth curve of a Ralstonia solanacearum culture over a period of 50 h. OD = optical density. Transition from lag- to log-phase not specifically shown.*

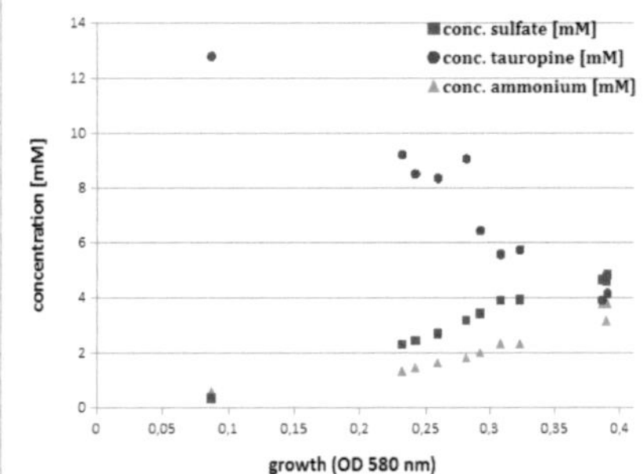

Figure 3: *Concentrations the substrate tauropine and the metabolites sulfate and ammonium in relation to the optical density. Values for optical density are equal to **Figure 2**.*

wards reaction were investigated.

Investigation of the tauropine dehydrogenase

The correct parameters of the assays, regarding the conditions and the fraction used, had to be established first. As it was not known yet, whether the tauropine dehydrogenase is membrane-associated or soluble, the three different fractions described above were investigated separately. The assay itself was modified using several buffers (CAPS, TRIS, KP, MOPS, MES) with pH values between 9.4 and 5.0. The electron acceptor was varied as well, using natural ones (cytochrome c, NAD$^+$, FMN), and artificial ones (2,6-Dichlorophenolindophenol, abbreviated DCPIP). The most promising conditions appeared to be MES/NaOH with pH=5 as buffer and DCPIP as an electron acceptor.

Table 1 shows the results of all enzyme assays regarding the tauropine dehydrogenase. It was induced only in the cultures grown with tauropine as carbon source. When taurine or acetate were used as carbon sources, no enzymatic activity could be determined, except in the soluble fraction II. This might be considered as false positive result, because of a high endogenous activity. In general, the enzyme activities are always low, which might be partially owed to the low amount of protein within the fractions. The measured activities have a high standard deviation as well, which leads to an imprecise determination of the activity of the potential tauropine dehydrogenase. It was not possible to determine, whether the enzyme is membrane-bound or not, because every fraction shows enzymatic activity. But the data strongly supports the assumption of the presence of a tauropine dehydrogenase.

Substrate for growth	Fraction	Average activity [mU/mg]	Standard deviation	Total amount of protein [mg/mL]
Tauropine	cell-free I	2.3	0.30	14.4
	cell-free II	3.0	0.89	6.3
	membrane II	3.9	2.14	4.4
	soluble II	1.3	0.08	4.7
	cell-free III	4.8	0.51	4.3
Taurine	cell-free I	0	–	2.6
	cell-free II	0	–	5.5
	membrane II	0	–	3.7
	soluble II	1.3	0.32	4.5
Acetate	Cell-free I	0	-	1.8

Table 1: Overview of tauropine dehydrogenase assays. Roman numerals indicate replicates from different days.

Investigation of the downstream pathway: the taurine dehydrogenase

After the existence of a tauropine dehydrogenase was proven, the enzymes involved in the pathway downstream were tested for their existence and activity. The next enzyme in the cascade is the taurine dehydrogenase. The setup of the assay was nearly equal to the previously described tauropine dehydrogenase assay. After testing different parameters, the most promising seemed to be a TRIS/HCl buffer with a pH=9 and cytochrome c as electron acceptor. Upon investigating the cell-free fraction II, the taurine dehydrogenase was found to be induced in the cultures grown with tauropine (activity: 12.1 mU/mg, standard deviation: 0.01) and taurine (activity: 27.2 mU/mg, standard deviation: 0.03). The fractions obtained from the cultures grown with acetate showed no activity. This proves the assumption, that the taurine dehydrogenase is rather induced, than constitutively expressed. Investigation of the mem-

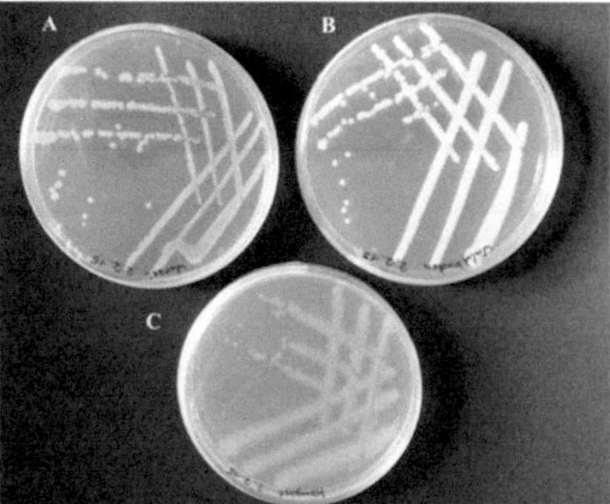

Figure 6: Environmental isolates grown on LB plates. A: root stock isolate; B: forest soil isolate; C: compost isolate

brane fraction II showed activity of the enzyme as well, exhibiting that the enzyme is membrane associated, as shown in **Figure 1**.

Investigation of the sulfoacetaldehyde acetyltransferase

The product of the taurine dehydrogenase, sulfoacetaldehyde, is further converted to acetylphosphate and hydrogen sulfite by the sulfoacetaldehyde acetyltransferase. The sulfoatetaldehyde acetyltransferase was assayed via a discontinuous measurement of the cell-free fraction I. In this assay the increase of acetyl phosphate was measured, using an iron(III)chelate as marker. Corresponding to the previous results, only the cultures grown with taurine and tauropine as sole carbon source showed enzyme activity. The average activities were 44.0 mU/mg and 34.2 mU/mg, respectively. It was not determined, whether the sulfoacetaldehyde acetyltransferase is membrane bound, because the activity was not yet assessed in the membrane and soluble fraction separately.

Investigation of the sulfite dehydrogenase

The sulfoacetaldehyde acetyltransferase-catalyzed reaction leads to hydrogen sulfite, which is further converted into sulfate by the sulfite dehydrogenase. The presence of this enzyme in all of the three cultures is reasonable, because the sulfate levels in the cultures were used for monitoring the growth rate (**Figure 3**). The activity of the sulfite dehydrogenase was assayed photometrically using the reduction of Fe(III) to Fe(II) as a marker. Upon reduction, the color of the solution changes. The enzymatic activity was strong in every culture, regardless the carbon source during growth. The values are 134.3 mU/mg, 87.8 mU/mg, and 102.4 mU/mg for the tauropine, taurine, and acetate grown cultures, respectively.

Alternative degradation pathway of taurine

An alternative dissimilation pathway of taurine to sulfoacetaldehyde could proceed via a taurine:pyruvate aminotransferase and an alanine dehydrogenase for pyruvate regeneration (enzymes not shown in **Figure 1**). To exclude this pathway, the alanine dehydrogenase was investigated in a photometric assay. The alanine dehydrogenase catalyzes an oxidative deamination. NAD^+ is reduced to NADH during the reactions, which leads to a change of the color of the solution. None of the three cultures exhibited the presence of an alanine dehydrogenase, which proved, that this pathway is not present. Therefore, the pathway depicted above (via a taurine dehydrogenase) is further confirmed.

All the enzyme assays carried out approved the suggested dissimilation pathway. To further identify and characterize the tauropine dehydrogenase, the different cultures were submitted to a 1D SDS-PAGE. The gel can be seen in **Figure 4**. Three bands were found to be induced and very strong (Spot A, B, C). They were selected for further peptide fingerprint mass spectrometry. Spot A from membrane fraction II was chosen, because the band was strong and only present in the membrane fraction of the tauropine grown culture. Spot B from the cell-free fraction III was chosen, because the band is found in every fraction of the tauropine grown culture and therefore might be the tauropine dehydrogenase. Spot C was suspected to be the sulfoacetaldehyde acetyltransferase, because the mass of nearly 66 kDa was already described by Ruff et al.[13] and it was induced in every culture, except the acetate grown one. The three spots were cut out, further processed and submitted to peptide fingerprinting-mass spectrometry (PF-MS). The results of the PF-MS analysis were used for a Mascot search. With scores of 788 (spot A), 720 (spot B), and 541 (spot C) the results are reliable. Spot A has a molecular weight of 80.2 kDa and was shown to be a TonB-dependent receptor in *Ralstonia*. It is known to be situated at the outer membrane and to transmit signals from the outside of the bacteria into the cytoplasm. The corresponding gene locus is MW7_0765. Directly adjacent, at position MW7_0766, a sulfite reductase is encoded. The sulfite reductase is expressed in the suggested degradation of tauropine and the TonB-dependent receptor might be induced as well, because of the proximity. Spot B has a molecular mass of 43 kDa and is the fragment of an elongation factor called Tu. The respective gene locus is MW7_3226. This protein is found in every fraction but in the membrane fraction. Spot C has a molecular weight of 65.7 kDa and corresponds to the sulfoacetaldehyde acetyltransferase enzyme. The gene locus is MW7_0623. This band was only present in the tauropine and taurine grown cul-

tures, because the acetate grown culture does not need the enzyme for substrate dissimilation.
Altogether, the tauropine dehydrogenase could not be found with the gel electrophoresis.

Tauropine was used as a carbon source in the cultures of different strains, as well as for the enzyme assays of the tauropine dehydrogenase. It is not commercially available and had to be synthesized according to the protocol of Gäde et al.[4] and Abderhalden et al.[15] The resulting tauropine product was purified by RP-HPLC. This method works well for removing side products, such as N-sulfoethane tauropine. But under the applied conditions, the seperation of tauropine from the starting material D-alanine was not possible. Using liquid extraction (butanol/water, pH=1), the yield of tauropine was increased, but D-alanine could not be removed completely. This was determined by NMR and MS analysis. NMR data is shown in the appendix. The concentration of the final aqueous solution of tauropine was determined to 0.75 M, but still some D-alanine remained in the solution. Consequently, some calculations based on this concentration may be inaccurate. The enzyme tests were not affected, because this freshly synthesized tauropine was not used for the assays. The analytical separation of D-alanine and tauropine is possible by acidifying the solution with sulfuric acid, using a 75 to 40 % gradient in HPLC.

We were not only interested in clarifying the dissimilation pathway of tauropine, but tried to find further potential tauropine degrading strains. New strains of tauropine degraders could facilitate the investigations on the degradation pathway stated in this work or even show the (additional) presence of another pathway. For isolation of new tauropine degrading bacterial strains, samples from different places were chosen: a forest soil sample, a sample from a root stock, and a compost sample were incubated in tauropine-mineral-salts medium. After several steps of selection, one strain of each source was selected and characterized. Microscopically, the bacteria isolated from the forest soil were sphere-shaped, small, and uniform, compared to the other selected strains. They did not exhibit much movement. The strain cultured from the root stock is roundish, immobile and seems to possess an envelop. The isolated strain from the compost was rod-shaped, thin and short, with a very mobile appearance. As expected, the environmental isolates also differ from each other on the macroscopic scale (**Figure 6**). The high mobility of the bacteria isolated from the compost leads to frayed colonies (sample C in **Figure 6**). On LB-plates, the strain from forest soil has a slimy appearance.
After the selection of one strain of each sample, the sulfate levels were determined to ensure, that the

strains are tauropine degrading. The sulfate concentration was as high as expected, upon using the previously synthesized tauropine. When the newly synthesized tauropine was used, the sulfate levels were markedly lower. This represents, as stated before, that the new tauropine solution still contains D-alanine. But it is also possible, that they are capable of using D-alanine as carbon source rather than tauropine. Thus the strains may depend on tauropine facultatively.

After it was sure, that the selected strains can degrade tauropine, the genus and species were determined by 16S rDNA sequence analysis. The DNAs of the three isolated strains and of the strain *Ralstonia solanacearum* were isolated and the 16S rDNA gene was multiplied via PCR. Additionally, the bacterial cells of *Ralstonia solanacearum* were lysed and submitted to PCR without previous isolation of the DNA. The products were applied to an agarose gel, depicted in **Figure 5**. Lane 4 and 5, counted from the left, show no band, indicating that the PCR has not worked properly. In this cases, the DNA has not been isolated before. The other bands were cut out and submitted to sequencing using universal 8f or 1492r primer. The sequences can be found in the appendix. By blast search, the genus could be identified, but not the exact species. The strain isolated from the root stock could not be assigned, because the sequencing

provided only a sequence of 44 nucleotides in length. The strains from the compost and forest soil were assigned to *Pseudomonas* on the basis of sequences of 232 and 410 nucleotides in length, respectively. The strain that was used for investigation on the tauropine degradation pathway was confirmed to be *Ralstonia solanacearum*, based on sequences of an average of 520 nucleotides in length. The sequence covering was between 99-100 %, because the strains are related to each other.

The three isolated strains and the strain *Ralstonia solanacearum* were tested for their capability to inhibit the growth of *Bacillus subtilis, Fusarium equiseti*, as well as each other. No effect on the growth of *Bacillus subtilis* and *Fusarium* could be obtained. Among themselves, the strains did not inhibit each other.

Summary and Outlook

In this work, the proposed degradation pathway of tauropine in *Ralstonia solanacearum* could partially be confirmed. The presence of a tauropine dehydrogenase was validated, even so its activity, as investigated in enzyme assays, was low and the measured data showed a high standard deviation. The separation of the proteins via gel electrophoresis did not help to identify the tauropine dehydrogenase. All possible enzyme activities downstream of the tau-

ropine dehydrogenase, suggested by previous findings in other organisms, could be confirmed. The lack of an alanine dehydrogenase confirmed the conversion of taurine to sulfoacetaldehyde via a taurine dehydrogenase. The existence of the sulfoacetaldehyde acetyltransferase was proven by PF-MS. In **Table 2** all tested enzymes and their respective activities are shown.

Enzyme	Tauropine grown	Taurine grown	Acetate grown
Tauropine dehydrogenase	2.3	0	0
Taurine dehydrogenase	12.1	27.3	0
Sulfoacetaldehyde acetyltransferase	34.2	44.0	0
Sulfite dehydrogenase	134.3	88.0	102.4
Alanine dehydrogenase	0	0	0

Table 2: *Overview of the results of all enzyme assays. Numbers refer to the activity of the respective enzyme in mU/mg.*

Furthermore, three new tauropine degrading strains could be selectively cultured, but not finally identified due to incomplete 16S rDNA sequencing. The strains originating from the compost and forest soil, could be assigned to the *Pseudomonas* family. The previous classification of model organism used in this study, *Ralstonia solanacearum*, was confirmed. The three environmental isolates and the strain *Ralstonia solanacearum* neither showed distinct antibacterial or antifungal activity against the test organism *Fusarium equiseti*, nor against each other. Purification of the synthesized tauropine proved to be challenging. Some D-alanine could not be separated from the tauropine.

The results of this work provide a good basis for further investigations. The purification method for tauropine could be refined to get pure substrate for new enzyme assays to determine, whether the tauropine dehydrogenase is membrane associated or not. To determine the mass of the tauropine dehydrogenase the cell lysates can be submitted to a 2D SDS-PAGE to get a better separation of the proteins.

Methods

Synthesis, purification, identification, and quantification of tauropine

Synthesis was conducted, as described already in 1931[4,15]. D-alanine (2.8 g) was mixed with 2-bromo ethane sulfonic acid (1,98 g) and 2 M sodium hydroxide solution (100 ml), and refluxed at 100°C overnight. The next day, a colorless solution was obtained. It was quenched with 2 M hydrochloric acid to pH=2. A white precipitate was obtained by vacuum evaporation. It consists of salty by-products and was removed by filtration.

The supernatant contained tauropine. For purification of tauropine, the supernatant was applied to a reversed phase 18 column, which was equilibrated with acetonitrile. Elution was started with dH_2O/acetonitrile (1/19). To elute tauropine, the percentage of water was slowly increased. All obtained fractions were vacuum evaporated. The resulting concentrated solutions were screened for their tauropine content by thin layer chromatography, using methanol/ethyl acetate (4/1) plus six drops of concentrated acetic acid as mobile phase, and 15 mM ninhydrin solution for staining. Tauropine was found by high performance liquid chromatography (HPLC) in that fraction, which had been eluted with dH_2O/acetonitrile (3/7). The dried tauropine product was diluted in dH_2O, sterile filtered, and the total yield was estimated by HPLC as 17.25 mM, using a 1 mM tauropine standard solution as reference.

For all HPLC measurements a hydrophilic interaction chromatography column (HILIC) was used, and 0.1 M ammonium acetate in acetonitrile/water (1/9), and pure acetonitrile, were used as eluents. A 75% to 60% gradient was utilized.

The product was identified by ^1H and ^{13}C nuclear magnetic resonance (NMR) analysis, using deuterium oxide as solvent.

^{1}H-NMR: (400.00 MHz, D_2O) δ_H: 3.81 (1H, q, J = 7.2 Hz), 3.50 (2H, m), 3.34 (2H, t, J = 6,8 Hz), 1.55 (3H, d, J = 7.2 Hz)

^{13}C-NMR: (400.00 MHz, D20) δ_C: 174.5 (C-2), 58.0 (C-3, 46.7 (C-7), 41.6 (C-6), 14.7 (C-12)

To test, whether the synthesized tauropine was free from toxins, a growth test in bacteria was performed.

Isolation and cultivation of tauropine-degrading strains from soil

Three samples of approximately 1 g of soil were collected in Konstanz, Germany, from forest soil, from the root of a dead standing tree, and from compost supernatant, respectively. Each sample was diluted in 3 ml of sterile culture medium (50 mM KH_2PO_4/K_2HPO_4, 0.25 mM $MgSO_4$, traces; for details see supporting information) with 6 mM tauropine, vortexed and agitated at 28°C over night. The next day, after soil particulates had settled, 50 µl of bacteria suspension were inoculated to fresh tauropine-selective culture medium. After another 24 and 48 h, another 50 µl of bacteria suspension were transferred to fresh liquid medium in order to select for the fastest strain to grow. Then each of the isolates was streaked on LB agar plates (5 g NaCl, 5 g tryptone, 2.5 g yeast extract, and 7.5 g agar per 500 ml of dH$_2$0). From the resulting overnight culture, one single culture was picked and transferred to tauropine-selective liquid medium. After another round of selection on plate and inoculation in liquid medium, the final pure isolates were obtained.

After each transfer, the progress of isolation was controlled by microscopy, using a magnification of 600x. Hereby the purity, outer appearance, size and motility of the bacteria were checked.

The isolated strains were provisionally named "forest soil", "root stock", and "compost", and together, they are referred to as "environment isolates".

16S rDNA analysis of the model organisms

DNA from the three environment isolates and from *Ralstonia solanacearum* was extracted by ethanolic precipitation (see supporting information). The resulting DNA pellet was diluted in 15 µL dH$_2$0, and DNA purity and concentration were measured by nanodrop spectrophotometry. The values obtained for DNA concentration at 260 nm were (in ng/µl): 838 (root stock), 507 (forest soil), 3026 (compost), and 136 (*Ralstonia solanacearum*). DNA purity was indicated by values 1,7<A260/A280<1,9.

For polymerase chain reaction (PCR), 75 ng of DNA templates were mixed in 0.2 ml tubes with Master Mix (ingredients and volumes see supporting information) and dH$_2$0 to a final volume of 49 µl. Then 1 µl of Phusion Hot Start II DNA Polymerase (2U/µl) was carefully stirred into the mixture, and the solution was centrifuged. A polymerase-matching thermocycler program was run (for details see supporting information). Afterwards, 8 µl of 6x loading buffer were mixed with each of the samples, and 29 µl of this mixture were applied to a 1% agarose gel (0.6 g agarose, 60 mL TAE-buffer (40 mM TRIS, 20 mM acetic acid, 1 mM EDTA), and 1 µl of ethidium bromide). Additionally, two pockets per gel were filled with 2 µl of a nucleotide molecular weight marker (1 kb DNA Ladder; Gene Ruler). The gel was then run at 120 Volt for 40 min. The DNA bands were visualized under UV light, and the DNA fragments indicating the 16S rDNA were accurately cut out using a scalpel. The DNA was isolated from the gel fragments using the Thermo Scientific "Silica Bead DNA Gel Extraction Kit", and the DNA was eluted from the silica gel with 10 µl of dH2O. Resulting DNA concentrations were (in ng/µl): 54.8 (root stock), 92.9 (forest soil), 61.9 (compost), and 97.6 (*Ralstonia solanacearum*). For DNA sequencing, 4 µl of DNA were carefully mixed with 3 µl of forward or reverse primer to a final primer concentration of 2 pg/µl, and with 8 µl

dH$_2$0. The cups were sent to Eurofins Genomics. Obtained nucleotide sequences were submitted to the NCBI nucleotide blast server at (http://www.ncbi.nlm.nih.gov/), which produced a list of closest known relatives, dependent on sequence similarity.

Screening for antibacterial and antifungal activity
Tauropine degrading strains, isolated as described above, were tested for their antibacterial or antifungal on *Bacillus subtilis* and *Fusarium equiseti* in bioassays. Suspension of one of the environment strains at a time was spotted in holes onto a lawn of growing *B. subtilis* and *Fusarium equiseti* in SFM plates (20 g/L soya flour, 20 g/L D-mannitol, agar-agar (Kobe I), dH$_2$0), respectively. The three isolated strains, and *Ralstonia solanacearum*, were also tested against each other on LB agar plates. After two days of incubation at 28°C, potential growth inhibition would have been determined by visible clearing zones.

Growth curves
12 ml of culture medium containing 8 mM tauropine were inoculated with 1% of *Ralstonia solanacearum* preparatory culture. Samples of 0.7 ml were taken at intervals. They were filled in a plastic cuvette to determine the optical density (OD) at 580 nm against air. The bacteria suspension from the cuvette was then centrifuged at 13.000 rpm for three minutes, and the precipitate was used to measure sulfate, ammonium, and tauropine concentrations. Sulfate concentrations were determined by OD measurement of barium sulfate precipitation at 436 nm (for details see supporting information). Ammonium concentration was assayed by the reaction of hypochlorite ions with sodium nitroprusside and ammonium in the dark, which generates a green dye. The intensity of this dye can be determined by OD measurement at 655 nm (for details see supporting information). Tauropine (retention time: 9.2 min) was quantified by HPLC in relation to a standard solution of known tauropine concentration.

Preparation of cell-free extract
Three 100 ml cell culture assays were prepared, containing culture medium, 0.7 % of *Ralstonia solanacearum* suspension, and 8 mM tauropine, 20 mM taurine, or a mixture of 20 mM acetate and 20 mM ammonium chloride. The exponential growth phase was identified by OD measurement at 580 nm, and cells were harvested at OD$_{580}$ values between 0.3 and 0.4. Therefore, each assay was centrifuged at 9.600 rpm for 15 minutes, before the pellet was washed once with washing buffer (50 mM K$_2$HPO$_4$/KH$_2$PO$_4$, 5 mM MgCl$_2$, pH=7.1). After centrifugation, as described above, the resulting cell pellet was collected in 3 mL of washing buffer. French Press, using 3 repeats and 1000 bar, was used for cell disruption. The resulting suspension was centrifuged for three minutes at 14.000 rpm and 6°C,

and the cell-free supernatant was used for enzymatic activity tests.

To separate particulate matter, ultracentrifugation (one hour at 4°C and 50.000 rpm) was used. The resulting pellet, which contained the cell membranes, was resuspended in 200 μl of washing buffer. Both the soluble fraction (the supernatant) and the membrane fraction (the pellet) were stored on ice until representative fractions were used for tauropine dehydrogenase assays.

The protein concentration of each extract was estimated by the Bradford method, which uses protein staining by Coomassie Brilliant Blue G250 (for details see supporting information).

Cell-free extracts, soluble fractions, or membrane fractions, that were obtained as described above, were used to determine the activities of the sulfite dehydrogenase, the tauropine dehydrogenase, the taurine dehydrogenase, the sulfoacetaldehyde acetyltransferase, and the alanine dehydrogenase.

For the sulfite dehydrogenase, oxidation of sulfite to sulfate was followed spectrophotometrically at 420 nm by reduction of Fe^{3+} to Fe^{2+}. The reaction mixture contained 950 μl of 50 mM TRIS-HCl pH=8, 10 μl of 100 mM potassium ferricyanide, 10 μl of 0.2 M sodium sulfite, and 30 μl of cell-free extract.

For measurements of the tauropine dehydrogenase in microorganisms no ideal reaction circumstances were known yet. The following reaction mixture showed the most significant activities: 950 μl MES/NaOH pH=5 + 10 μl dichlorophenolindophenol (DCPIP) + 50 μl extract + 50 μl 320 mM tauropine. Reduction of DCPIP was followed at 600 nm.

The tauropine backwards reaction, where the formation of tauropine from pyruvate and taurine is followed, was conducted as following: 800 μl of TRIS pH=8 were mixed with 0.5 mM NADH in TRIS, 50 mM taurine, 2.5 mM pyruvate, and cell-extract, and extinction was measured at 365 nm.

For taurine dehydrogenase activity assay, the method for the tauropine dehydrogenase was changed insofar, as tauropine solution was exchanged by taurine solution.

For the sulfoacetaldehyde acetyltransferase, the increase of acetyl phosphate was discontinuously measured by a colorimetric assay of iron(III)chelate (for details see supporting information).

For measurements of the alanine dehydrogenase, which catalyzes an oxidative deamination reaction, formation of NADH out of NAD^+ was monitored at 365 nm (for details see supporting information).

A 13 % separation SDS-PAGE-gel was layered with a 6 % collecting gel (for detailed information accord-ing the ingredients of the solutions see supporting information). After successful polymerization, every single gel pocket was carefully rinsed with TRIS-glycin-SDS running buffer. The samples, that contained 50 μg of protein from cell-free extracts, were diluted with dH_2O to a final volume of 30 μl, mixed with 10 μl of protein, and heated for six minutes at 99°C. The pockets were then filled with the protein samples or a molecular protein weight marker. The gel was run at 33 mA. The proteins were stained with triphenyl methane dye (Coomassie Blue, for details see supporting information) overnight, and afterwards washed with 200 mM acetic acid in dH_2O/methanol (1/1). Protein band patterns were compared, and interesting bands at 45, 60 and 90 kDa were cut out accurately with a scalpel. They were submitted to peptide fingerprinting mass spectrometry (PF-MS) for analysis with matrix assisted laser desorption ionization - time of flight (MALDI-TOF). For storage, the gel was vacuum dried between two cellophane sheets.

1   Morizawa, K. Über die Extraktstoffe von *Octopus octopodia*. *Acta School of Medical University Kyoto* **10**, 285-298 (1928).

2   Pörtner, H. O. Environmental and functional limits to muscular exercise and body size in marine invertebrate athletes. *Comparative Biochemistry and Physiology Part A: Molecular & Integrative Physiology* **133**, 303-321 (2002).

3   Sato, M., Takeuchi, M., Kanno, N., Nagahisa, E. & Sato, Y. Distribution of opine dehydrogenases and lactate dehydrogenase activities in marine animals. *Comparative Biochemistry and Physiology Part B: Comparative Biochemistry* **106**, 955-960 (1993).

4   Gäde, G. Purification and properties of tauropine dehydrogenase from the shell adductor muscle of the ormer, *Haliotis lamellosa*. *European Journal of Biochemistry* **160**, 311-318 (1986).

5   Kanno, N., Sato, M., Nagahisa, E. & Sato, Y. Tauropine dehydrogenase from the sandworm *Arabella iricolor* (Polychaeta: Errantia): Purification and characterization. *Comparative Biochemistry and Physiology Part B: Biochemistry and Molecular Biology* **114**, 409-416 (1996).

6   Kan-no, N., Sato, M., Yokoyama, T., Nagahisa, E. & Sato, Y. Tauropine dehydrogenase from the starfish *Asterina pectinifera* (Echinodermata: Asteroidea): presence of opine production pathway in a deuterostome invertebrate. *Comparative Biochemistry and Physiology Part B: Biochemistry and Molecular Biology* **121**, 323-332 (1998).

7   Kan-no, N., Matsu-ura, H., Jikihara, S., Ya-mamoto, B., Endo, N., Moriyama, S., Nagahisa, E. & Sato, M. Tauropine dehydrogenase from the marine sponge *Halichondria japonica* is a homolog of ornithine cyclodeaminase/mu-crystallin. *Comparative Biochemistry and Physiology Part B: Biochemistry and Molecular Biology* **141**, 331-339 (2005).

8   Sato, M., Takeuchi, M., Kanno, N., Nagahisa, E. & Sato, Y. Purification and properties of tauropine dehydrogenase from a red alga, crystallin. *Comparative Biochemistry and*

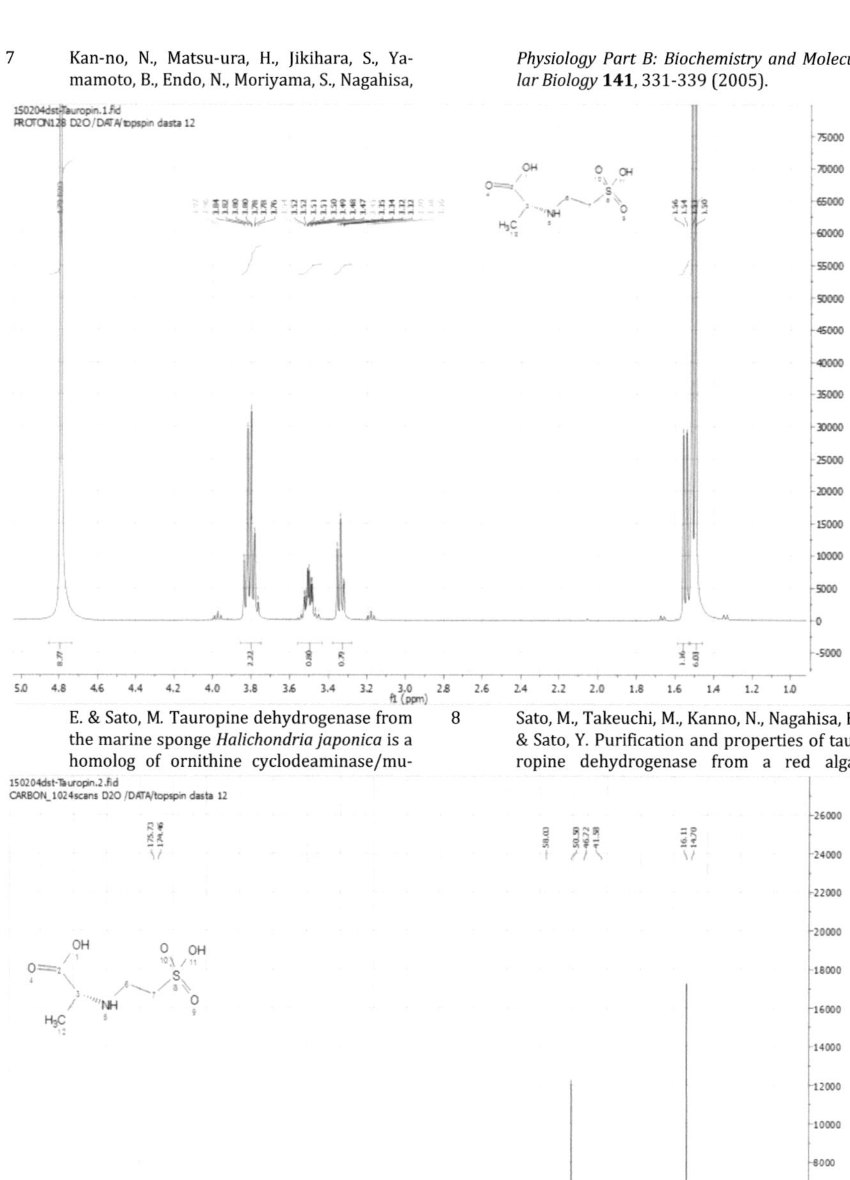

Rhodoglossum japonicum. *Hydrobiologia* **260/261**, 673-678 (1993).

9   Lang, J., Planamente, S., Mondy, S., Dessaux, Y., Morera, S. & Faure, D. Concerted transfer of the virulence Ti plasmid and companion At plasmid in the *Agrobacterium tumefaciens*-induced plant tumour. *Molecular Microbiology* **90**, 1178-1189 (2013).

10 Flores-Mireles, A. L., Eberhard, A. & Winans, S. C. *Agrobacterium tumefaciens* can obtain sulphur from an opine that is synthesized by octopine synthase using S-methylmethionine as a substrate. *Molecular Microbiology* **84**, 845-856 (2012).

11 Moore, L. W., Chilton, W. S. & Canfield, M. L. Diversity of opines and opine-catabolizing bacteria isolated from naturally occurring crown gall tumors. *Applied and Environmental Microbiology* **63**, 201-207 (1997).

12 Denger, K., Smits, T. & Cook, A. L-Cysteate sulpho-lyase, a widespread pyridoxal 5'-phosphate-coupled desulphonative enzyme purified from *Silicibacter pomeroyi DSS-3T*. *Biochemical Journal* **394**, 657-664 (2006).

13 Ruff, J., Denger, K. & Cook A. Sulphoacetaldehyde acetyltransferase yields acetyl phosphate: purification from *Alcaligenes defragrans* and gene clusters in taurine degradation. *Biochemical Journal* **369**, 275-285 (2003).

14 Brüggemann, C., Denger, K., Cook, A. M. & Ruff, J. Enzymes and genes of taurine and isethionate dissimilation in *Paracoccus denitrificans. Microbiology* **150**, 805-816 (2004).

15 Abderhalden, E. & Haase, E. Gewinnung von Iminodicarbonsäuren aus Aminosäuren und Halogenfettsäuren. *Hoppe-Seyler's Zeitschrift für physiologische Chemie* **202**, 49-55 (1931).

Appendix

Nucleotide sequences

Compost – 1492r primer: 183-484 of sequence
CTATCGGTTTTATGGGATTAGCTCCACCTCGCGGCTTG
GCAACCCTTTGTACCGACCATTGTAGCACGTGTGTAGC
CCTGGCCGTAAGGGCCATGATGACTTGACGTCATCCCC
ACCTTCCTCCGGTTTGTCACCGGCAGTCTCCTTAGAGT
GCCCACCATAACGTGCTGGTAACTAAGGACAAGGGTT
GCGCTCGTTACGGGACTTAACCCAACATCTCACGACAC
GAGCTGACGACAGCCATGCAGCACCTGTGTCTGAGCTC
CCGAAGGCACCAATCCATCTCTGGAAAGTTCTCAGCA

Compost – 8f primer: 149-315 of sequence
AAGCAGGGGACCTTCGGGCCTTGCGCTATCCGATGAGC
CTAGGTCGGATTAGCTTGTTGGTGAGGTAATGGCTCA
CCAAGGCGACGATCCGTAACTGGTCTGAGAGGATGAT
CAGTCACACTGGAACTGAAACACGGTCCAGACTCCTAC
GGGAGGCAGCAGTGGGG

Ralstonia solanacearum – 1492r primer: 158-752 of sequence
AGTTGCAGACTACGATCCGGACTACGATGCATTTTCTG
GGATTAGCTCCACCTCGCGGCTTGGCAACCCTCTGTAT
GCACCATTGTATGACGTGTGAAGCCCTACCCATAAGG
GCCATGAGGACTTGACGTCATCCCCACCTTCCTCCGGT
TTGTCACCGGCAGTCTCTCTAGAGTGCTCTTGCGTAGC
AACTAAAGACAAGGGTTGCGCTCGTTGCGGGACTTAA
CCCAACATCTCACGACACGAGCTGACGACAGCCATGCA
GCACCTGTGTCCACTTTCCCTTTCGGGCACCTAATGCA
TCTCTGCTTCGTTAGTGGCATGTCAAGGGTAGGTAAG
GTTTTTCGCGTTGCATCGAATTAATCCACATCATCCAC
CGCTTGTGCGGGTCCCCGTCAATTCCTTTGAGTTTTAA
TCTTGCGACCGTACTCCCCAGGCGGTCAACTTCATGCG
TTAGCTTCGTTACTAAGGAAATGAATCCCCAACAACT
AGTTGACATCGTTTAGGGCGTGGACTACCAGGGTATC
TAATCCTGTTTGCTCCCCACGCTTTCGTGCATGAGCGT
CAGTGTTATCCCAGGGGGCTGCCTTTCGCC

Ralstonia solanacearum – 8f primer: 95-540 of sequence
CCTGTAGTGGGGGATAACTAGTCGAAAGATTACCTAA
TACCGCATACGACCTGAGGGTGAAAGTGGGGGACCGC
AAGGCCTCACGCTATAGGAGCGGCCGATATCTGATTA
GCTAGTTGGTGGGGTAAAGGCCTACCAAGGCGACGAT
CAGTAGCTGGTCTGAGAGGACGATCAGCCACACTGGG
ACTGAGACACGGCCCAGACTCCTACGGGAGGCAGCAGT
GGGGAATTTTGGACAATGGGCGAAAGCCTGATCCAGC
AATGCCGCGTGTGTGAAGAAGGCCTTCGGGTTGTAAA
GCACTTTTGTCCGGAAAGAAATCCCTTGGGATAATAC
CTCGGGGGGGATGACGGTACCGGAAGAATAAGGACCGG
CTAACTACGTGCCAGCAGCCGCGGTAATAGTAGGGTCC
GAGCGTTAATCGGAATTACTGGGCGTAAAGCGTGCG

Forest soil – 1492r primer: 178-682 of sequence
ACGATCGGTTTTGTGGGATTAGCTCCACCTCGCGGCTT
GGCAACCCTCTGTACCGACCATTGTAGCACGTGTGTAG
CCCAGGCCGTAAGGGCCATGATGACTTGACGTCATCCC
CACCTTCCTCCGGTTTGTCACCGGCAGTCTCCTTAGAG
TGCCCACCATTACGTGCTGGTAACTAAGGACAAGGGT
TGCGCTCGTTACGGGACTTAACCCAACATCTCACGACA
CGAGCTGACGACAGCCATGCAGCACCTGTCTCAATGTT
CCCGAAGGCACCAATCCATCTCTGGAAAGTTCATTGGA
TGTCAAGGCCTGGTAAGGTTCTTCGCGTTGCTTCAAAT
TAAACCACATGCTCCACCGCTTGTGCGGGCCCCGTCA
ATTCATTTGAGTTTTAACCTTGCGGCCGTACTCCCCAG

GCGGTCAACTTAATGCGTTAGCTGCGCCACTAAGAGCT
CAAGGCTCCCAACGGCTAGTTGACCATCGTTTACGGCG
TGGAACTACCAG

Forest soil – 8f primer: 133-448 of sequence
CATACGTCCTACGGGAGAAAGCAGGGGACCTTCGGGC
CTTGCGCTATCAGATGAGCCTAGGTCGGATTAGCTAG
TTGGTGAGGTAATGGCTCACCAAGGCGACGATCCGTA
ACTGGTCTGAGAGGATGATCAGTCACACTGGAACTGA
GACACGGTCCAGACTCCTACGGGAGGCAGCAGTGGGG
AATATTGGACAATGGGCGAAAGCCTGATCCAGCCATG
CCGCGTGTGTGAAGAAGGTCTTCGGATTGTAAAGCAC
TTTAAGTTGGGAGGAAGGGCATTAACCTAATACGTTA
GTGTTTTGACGTTACCGACA

Rootstock – 1492r primer: 0

Rootstock – 8f primer: 24-73 of sequence
ATGGCTCAGTAAGTCGTAACAAGGTAACCAAGTAGAG
TTTGATCATGGA

YOUR KNOWLEDGE HAS VALUE

- We will publish your bachelor's and master's thesis, essays and papers

- Your own eBook and book - sold worldwide in all relevant shops

- Earn money with each sale

Upload your text at www.GRIN.com and publish for free